Published 2012 by Cliff Top Press Ltd.
www.arithmeticvillage.com

ISBN 978-0-9845731-9-6

Printed by Lightning Source, USA

Files licensed by www.depositphotos.com: Leather © montego • Background of burlap hessian sacking © odua
• Cardboard background © Leonardi • harlequin pattern fabric background © ratselmeister • Purple squares © jele76

King David Divide

By Kimberly Moore

To Shel

King David Divide is thoughtful and wise,
sharing equally is what he decides.

If forty jewels arrive in his den,

four people want them,

they each leave with ten.

A juggling jester is next in line.

The only one there, the King gives him nine.

Twentyone jewels tumble from the sky.

Three happy friends catch seven passing by.

Thirty jewels float
around the kings moat,

five sailors catch six each, in every boat.

Two small children arrive at the king's gate.

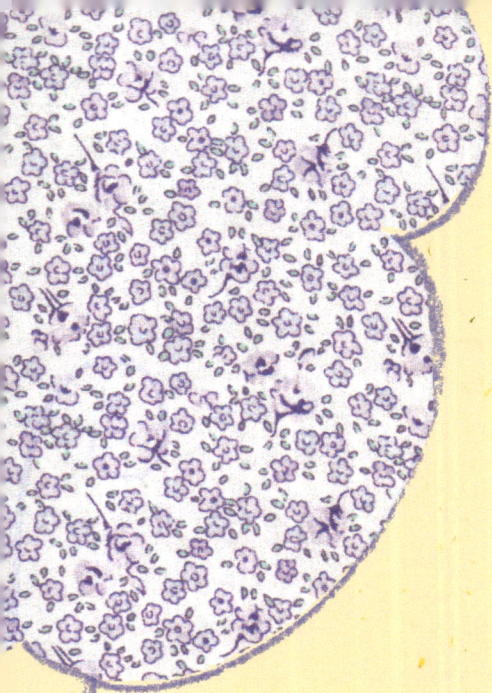

Sixteen gems to share, they both receive eight.

When he divides, if some are left over,

they go to his pet, a cute dog named Rover!

Eleven shiny jewels for three young boys,

the two extra gems are Rover's new toys.

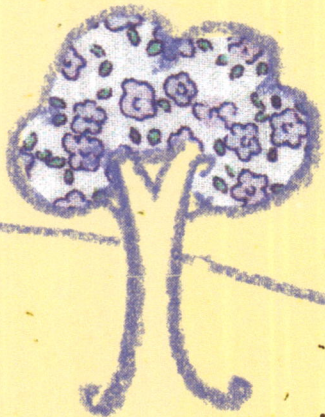

If you are a ruler who's kind and who's fair,

act like King David and learn how to share!

Dear Grownup,

King David Divide demonstrates graciously, the concept of division.

While reading this book, some children may want to follow the division equation represented in each picture. Others may wish to act like King David and share the jewels. Remember to provide a 'Rover's bowl' in case there are some left over.

Seek out opportunities in everyday life to share equitably. A light hearted approach to learning, can support a lifetime love of math!

For more inspirational ideas visit www.arithmeticvillage.com

In undivided service,

Kimberly

Arithmetic Village

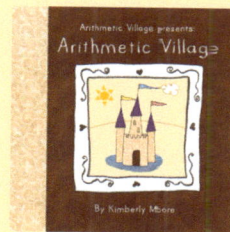

Polly Plus

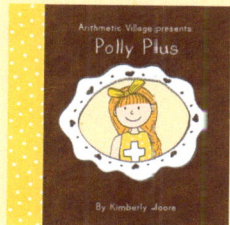

Linus Minus

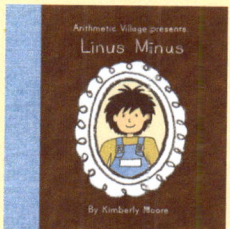

Tina Times

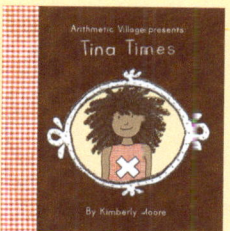

King David Divide

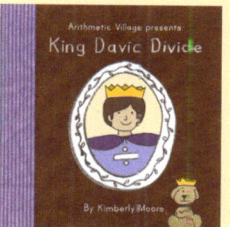